Chapter I. Introduction
A. The significance of Vedic Mathematics in ancient Indian culture
B. Explaining the purpose and benefits of studying Vedic Mathematics today

Chapter II. Origins and Philosophy of Vedic Mathematics
A. Historical background of Vedic Mathematics
B. Understanding the philosophical foundations of Vedic Mathematics
C. The connection between Vedic Mathematics and Vedic literature

Chapter III. The Sutras: Fundamental Principles of Vedic Mathematics
A. Introduction to the 16 Sutras and their applications
B. Detailed explanation of each Sutra and its significance
C. Illustrative examples and step-by-step solutions for various mathematical operations

Chapter IV. Arithmetic Operations
A. Enhancing mental calculation through Vedic Mathematics
B. Addition and subtraction techniques using Vedic methods
C. Multiplication techniques: one-line and specific digit methods
D. Division techniques: duplex method and Nikhilam method

Chapter V. Practical Applications and Relevance
A. Real-world examples showcasing the practicality of Vedic Mathematics
B. Enhancing speed and accuracy in competitive exams
C. Incorporating Vedic Mathematics into daily life and professional fields

Chapter VI. Criticisms and Controversies
A. Addressing common criticisms against Vedic Mathematics
B. Debunking misconceptions and clarifying its limitations
C. Exploring alternative viewpoints and debates

Chapter VII. Future Prospects and Further Study

A. Emerging research and developments in Vedic Mathematics
B. Potential areas of growth and innovation

Chapter VIII. Conclusion
A. Summary of key takeaways from the book
B. Encouraging readers to delve deeper into Vedic Mathematics
C. Final thoughts on the enduring relevance of this ancient
mathematical system

Chapter I
Introduction to Vedic Mathematics

Mathematics has been an integral part of human civilization, facilitating scientific discoveries, technological advancements, and economic progress. Amidst the vast array of mathematical systems and methods developed throughout history, Vedic Mathematics stands out as a unique and ancient approach that originated in the Indian subcontinent.

Vedic Mathematics traces its roots to the ancient Indian civilization, particularly the Vedic period, which spanned from around 1500 BCE to 500 BCE. The term "Vedic" refers to the Vedas, a collection of sacred texts that form the foundation of Indian philosophy, spirituality, and knowledge. Within these ancient texts, profound insights into the nature of numbers and calculations were discovered and codified.

The significance of Vedic Mathematics lies not only in its practical applications but also in its profound philosophical underpinnings. The Vedic civilization recognized the fundamental interconnectedness of all things and sought to express this interconnectedness through mathematical principles. Vedic Mathematics embodies this holistic worldview, reflecting a deep understanding of the natural order and the underlying harmony of the universe.

The purpose of studying Vedic Mathematics today extends beyond its historical and cultural significance. It offers a fresh and alternative approach to mathematical calculations, fostering mental agility, critical thinking, and a deeper understanding of mathematical concepts. By exploring the techniques and methods of Vedic Mathematics, individuals can enhance their problem-solving abilities, improve their computational speed, and develop a more intuitive grasp of numbers.

One of the distinguishing features of Vedic Mathematics is its emphasis on mental calculation techniques. The ancient sages and scholars who developed Vedic Mathematics recognized the power of the human mind in performing calculations swiftly and accurately. Vedic techniques enable individuals to perform complex calculations mentally, often in a single line or with remarkable speed, reducing reliance on external aids such as calculators or written computations.

Vedic Mathematics encompasses a wide range of mathematical operations, including arithmetic, algebra, geometry, calculus, and more. It provides elegant and efficient methods for addition, subtraction, multiplication, division, factorization, solving equations, finding areas, volumes, and trigonometric values, and tackling advanced mathematical concepts.

Moreover, Vedic Mathematics offers practical applications beyond academic pursuits. Its techniques can be applied in everyday life, from managing personal finances and calculating bills to estimating quantities and making quick estimations. The versatility and adaptability of Vedic Mathematics make it a valuable tool for individuals in various fields, including engineering, finance, research, and any profession that involves numerical analysis.

In this book, we embark on a journey to explore the depths of Vedic Mathematics. Each chapter will delve into specific aspects of this ancient system, unraveling the underlying principles, elucidating the techniques, and providing illustrative examples. By the end of this book, readers will gain a comprehensive understanding of Vedic Mathematics and its potential to revolutionize the way we approach mathematical calculations.

Whether you are a student seeking to improve your mathematical skills, a teacher looking for innovative teaching methodologies, or simply an enthusiast intrigued by the wonders of ancient knowledge, this book will serve as a guide to unlock the secrets and mysteries of Vedic Mathematics.

So, let us embark on this enlightening journey together, as we unravel the beauty and power of Vedic Mathematics.

A. The Significance of Vedic Mathematics in Ancient Indian Culture:

Vedic Mathematics holds immense significance in ancient Indian culture, as it emerged from the profound spiritual and intellectual traditions of the Vedic civilization. The term "Vedic" refers to the Vedas, the ancient sacred texts of India, which are considered a treasure trove of knowledge covering various disciplines, including mathematics.

Integration of Mathematics and Spirituality: In ancient India, mathematics was not viewed as a separate and isolated subject but was deeply integrated with spirituality and cosmic principles. The sages and scholars of the Vedic era believed that the universe is inherently mathematical and sought to understand its underlying harmony through the study of numbers and calculations. Vedic Mathematics reflects this holistic worldview, as it incorporates spiritual insights and cosmic symbolism into its mathematical techniques.

Preservation of Knowledge: Vedic Mathematics played a vital role in preserving and transmitting mathematical knowledge across generations. The oral tradition of teaching and learning ensured the continuity of this ancient system, passing it down through the ages. The mnemonic devices and poetic formulations employed in Vedic Mathematics helped in memorizing and transmitting mathematical concepts effectively.

Efficiency and Mental Calculation: Vedic Mathematics is renowned for its emphasis on mental calculation techniques and shortcuts. Ancient Indians recognized the importance of mental agility and speed in mathematical computations. Vedic Mathematics provides efficient methods for performing complex calculations mentally, enabling individuals to solve problems swiftly and accurately without relying heavily on external aids like calculators or written calculations.

Versatility and Flexibility: Vedic Mathematics offers a versatile and flexible approach to problem-solving. It provides a diverse range of techniques and methods that can be applied across various mathematical domains, including arithmetic, algebra, geometry, and calculus. The adaptability of Vedic Mathematics allows individuals to approach problems from different angles and choose the most suitable method for solving them.

Cultivation of Intuition and Creativity: Vedic Mathematics encourages the development of intuition and creativity in mathematical thinking. Instead of relying solely on rote memorization and mechanical procedures, Vedic techniques stimulate individuals to think critically, make connections, and explore alternative approaches. This approach fosters a deeper understanding of mathematical concepts and cultivates problem-solving skills that extend beyond the realm of mathematics.

Connection with Cultural Identity: Vedic Mathematics forms an integral part of India's cultural identity and heritage. It represents a remarkable intellectual achievement of ancient Indian civilization and showcases the contributions made by Indian mathematicians and scholars. Studying Vedic Mathematics allows individuals to connect with their cultural roots and appreciate the intellectual and scientific advancements of ancient India.

In conclusion, Vedic Mathematics holds great significance in ancient Indian culture due to its integration of mathematics and spirituality, its role in preserving knowledge, its emphasis on mental calculation, its versatility, its cultivation of intuition and creativity, and its connection with India's cultural identity. Understanding and exploring Vedic Mathematics not only provides insights into the mathematical practices of ancient India but also offers a unique perspective on the interconnectedness of mathematics, spirituality, and human consciousness.

B. The Purpose and Benefits of Studying Vedic Mathematics Today:

Enhancing Mental Calculation Skills: One of the primary purposes of studying Vedic Mathematics is to improve mental calculation skills. Vedic techniques offer efficient methods for performing arithmetic operations mentally, enabling individuals to quickly and accurately solve complex calculations. By practicing Vedic Mathematics, individuals can enhance their speed and accuracy in mental calculations, making them more proficient in various mathematical tasks in everyday life.

Developing a Deeper Understanding of Mathematics: Studying Vedic Mathematics goes beyond memorizing formulas and procedures. It fosters a deeper understanding of mathematical concepts and principles. Vedic techniques often provide intuitive and alternative approaches to problem-solving, encouraging individuals to think critically and explore mathematical relationships. This deeper understanding not only enhances mathematical abilities but also cultivates a broader perspective on the subject.

Improving Problem-Solving Abilities: Vedic Mathematics equips individuals with a diverse range of problem-solving techniques. These methods offer alternative strategies for approaching mathematical problems, enabling individuals to find creative and efficient solutions. By incorporating Vedic techniques into problem-solving, individuals can develop their analytical thinking skills, logical reasoning, and adaptability when faced with mathematical challenges.

Speed and Efficiency in Competitive Exams: Vedic Mathematics is particularly valuable for individuals preparing for competitive exams, such as entrance exams, aptitude tests, or standardized tests. The mental calculation techniques and shortcut methods in Vedic Mathematics allow test-takers to solve problems more quickly, saving valuable time during exams. This speed and efficiency can give individuals a competitive edge in time-sensitive exams.

Holistic and Interconnected Approach: Vedic Mathematics offers a holistic and interconnected approach to mathematics. Rather than treating different mathematical topics as separate entities, Vedic Mathematics emphasizes the interrelations and common principles across various domains of mathematics. This holistic understanding helps individuals develop a comprehensive and integrated view of mathematics, allowing them to make connections between different concepts and apply their knowledge more effectively.

Cultivating Analytical and Critical Thinking Skills: Studying Vedic Mathematics nurtures analytical and critical thinking skills. Vedic techniques often require individuals to break down complex problems into simpler components and identify patterns and relationships. By practicing Vedic Mathematics, individuals develop the ability to analyze problems, think logically, and employ strategic thinking to arrive at solutions.

Practical Applications in Daily Life: Vedic Mathematics has practical applications beyond academic settings. The mental calculation techniques and shortcut methods can be applied to everyday situations such as calculating bills, estimating quantities, managing finances, or measuring dimensions. The efficiency gained from practicing Vedic Mathematics can save time and enhance accuracy in practical tasks, making it a valuable skill for various professions and day-to-day activities.

In summary, studying Vedic Mathematics today serves the purpose of enhancing mental calculation skills, deepening the understanding of mathematics, improving problem-solving abilities, facilitating speed and efficiency in competitive exams, promoting a holistic approach to mathematics, cultivating analytical and critical thinking skills, and offering practical applications in daily life. These benefits make Vedic Mathematics a valuable and relevant subject of study in the modern world.

Chapter II
The Origins and Philosophy of Vedic Mathematics:

Origins of Vedic Mathematics: Vedic Mathematics finds its origins in ancient India during the Vedic period, which dates back to approximately 1500 BCE to 500 BCE. The term "Vedic" refers to the Vedas, a collection of sacred texts considered the foundation of Indian knowledge, spirituality, and philosophy. Vedic Mathematics is believed to have been developed by ancient sages and scholars, who sought to understand and express the profound truths of the universe through the language of numbers.

The philosophy of Vedic Mathematics: Vedic Mathematics is rooted in the philosophical traditions of the Vedic civilization, which recognized the fundamental interconnectedness of all aspects of existence. The philosophy underlying Vedic Mathematics is deeply spiritual and embodies the holistic worldview of the Vedic texts.

Cosmic Consciousness and Unity: Vedic Mathematics views the universe as an interconnected web of consciousness, where all phenomena are interrelated. It sees mathematics as a language that helps us decipher and understand the underlying unity and harmony of the cosmos. In Vedic philosophy, the study of mathematics is not merely a means to solve practical problems but also a path to attaining higher states of consciousness and realizing the oneness of all existence.

The Interplay of Consciousness and Numbers: Vedic Mathematics recognizes that numbers are not mere abstract symbols but manifestations of consciousness itself. Each number holds profound significance and embodies certain qualities or energies. The study of numbers is seen as a means to unlock the deeper truths of existence and gain insights into the nature of reality.

Mental and Intuitive Approach: Vedic Mathematics emphasizes the power of the human mind and intuition in mathematical calculations. The ancient sages believed in the inherent ability of the human mind to perform complex calculations swiftly and accurately. Vedic techniques provide mental calculation strategies that rely on observation, pattern recognition, and intuitive reasoning. This approach encourages individuals to tap into their innate mental faculties and develop a deep understanding of numbers.

Simplicity and Elegance: Vedic Mathematics is characterized by its simplicity and elegance. The ancient sages sought to develop methods that were easy to understand, remember, and apply. The techniques are often derived from fundamental principles, and their simplicity allows for efficient mental calculations. The emphasis on simplicity reflects the belief that the laws governing the universe are inherently simple and accessible to all.

Wholeness and Integration: Vedic Mathematics integrates various branches of mathematics, including arithmetic, algebra, geometry, and calculus, into a unified system. The interconnectedness of mathematical concepts is highlighted, promoting a holistic understanding of mathematics. Vedic Mathematics aims to provide a comprehensive and integrated approach to problem-solving, encouraging individuals to explore connections between different branches of mathematics and develop a cohesive understanding of the subject.

In summary, the origins and philosophy of Vedic Mathematics can be traced back to ancient India during the Vedic period. Its philosophy is deeply rooted in the interconnectedness of the universe, the consciousness inherent in numbers, the power of the human mind, simplicity and elegance, and the integration of different branches of mathematics. Understanding the philosophy behind Vedic Mathematics not only enhances mathematical skills but also offers a broader perspective on the profound interplay between mathematics, consciousness, and the nature of reality.

A. Historical Background of Vedic Mathematics:

Vedic Mathematics finds its roots in ancient India during the Vedic period, which spans from approximately 1500 BCE to 500 BCE. The term "Vedic" refers to the Vedas, a collection of sacred texts that form the foundation of Indian philosophy, spirituality, and knowledge. Vedic Mathematics is believed to have been developed by ancient sages and scholars who were deeply immersed in the study of the Vedas and sought to unravel the mysteries of numbers and calculations.

Vedas and Mathematical Insights: The Vedas, which consist of four main texts (Rigveda, Samaveda, Yajurveda, and Atharvaveda), contain hymns, prayers, rituals, and philosophical teachings. Amidst this vast body of knowledge, there are mathematical insights and techniques embedded in various sections. The ancient sages recognized the significance of numbers and mathematics as tools for understanding the cosmos and expressing cosmic principles.

Oral Tradition and Transmission: During the Vedic period, knowledge was primarily transmitted through an oral tradition, where teachings were passed down from generation to generation through recitation and memorization. This method ensured the preservation and continuity of Vedic Mathematics, allowing it to survive over thousands of years.

Sutras and Formulations: The core principles of Vedic Mathematics are encapsulated in a set of concise and memorable statements called "Sutras." The term "Sutra" means "thread" or "formula" and serves as a guiding principle for problem-solving. These Sutras are concise aphorisms that contain mathematical techniques and strategies. Each Sutra represents a specific mathematical operation or concept and provides a framework for solving mathematical problems.

Contribution of Ancient Scholars: Several ancient scholars and mathematicians made significant contributions to the development and refinement of Vedic Mathematics. Scholars such as Bharati Krishna Tirthaji (1884-1960), who is often credited with reviving and popularizing Vedic Mathematics in the modern era, played a crucial role in compiling and organizing the Vedic techniques for wider dissemination.

Integration into Indian Mathematical Tradition: Vedic Mathematics is not an isolated mathematical system but is deeply integrated into the broader Indian mathematical tradition. Throughout history, India has been a hub of mathematical advancements, with scholars making significant contributions to various branches of mathematics. Vedic Mathematics seamlessly blends with other mathematical systems, providing additional tools and approaches to problem-solving.

Modern Revival and Recognition: In recent years, Vedic Mathematics has gained attention and recognition globally. The simplicity, efficiency, and mental calculation techniques of Vedic Mathematics have appealed to students, educators, and mathematicians alike. The system has been incorporated into educational curricula in some schools and has been the subject of research and exploration in the field of mathematics education.

In summary, Vedic Mathematics emerged during the Vedic period in ancient India, as an amalgamation of mathematical insights embedded in the Vedas. It was developed and transmitted through the oral tradition, encapsulated in concise Sutras, and further refined by ancient scholars. Today, Vedic Mathematics continues to be a vibrant mathematical system, blending with the broader Indian mathematical tradition and gaining recognition for its unique approaches to problem-solving.

B. Understanding the Philosophical Foundations of Vedic Mathematics:

Vedic Mathematics is not merely a set of techniques and methods; it is deeply rooted in the philosophical foundations of ancient Indian culture. The philosophical principles underlying Vedic Mathematics shape its approach, techniques, and understanding of numbers. By exploring these foundations, we can gain a deeper appreciation of the holistic and interconnected nature of Vedic Mathematics.

Unity and Interconnectedness: The philosophical foundation of Vedic Mathematics is rooted in the belief that everything in the universe is interconnected. It recognizes the inherent unity between numbers, the cosmos, and the human mind. Vedic Mathematics sees numbers as expressions of universal principles and consciousness, representing the interconnected fabric of reality. This interconnectedness is reflected in the techniques and methods of Vedic Mathematics, which emphasize the interrelations between different branches of mathematics and foster a holistic understanding of mathematical concepts.

Cosmic Harmony and Balance: Vedic Mathematics acknowledges the underlying harmony and balance that permeate the cosmos. It views mathematics as a means to access and express this cosmic harmony. The techniques of Vedic Mathematics are designed to align with this harmony, employing elegant and efficient methods that resonate with the natural order. By adhering to this principle of cosmic harmony, Vedic Mathematics promotes simplicity, elegance, and efficiency in problem-solving.

Intuition and Insight: Vedic Mathematics recognizes the power of intuition and insight in the pursuit of mathematical understanding. It encourages individuals to tap into their innate intuition and develop a deep connection with numbers. Rather than relying solely on mechanical procedures, Vedic Mathematics emphasizes the importance of intuitive reasoning, observation, and pattern recognition. By cultivating intuition, individuals can access deeper insights into mathematical concepts and discover alternative approaches to problem-solving.

Practicality and Applicability: The philosophical foundations of Vedic Mathematics prioritize practicality and applicability in real-world scenarios. Vedic Mathematics is not limited to abstract concepts and theoretical frameworks but offers practical techniques that can be readily applied to various mathematical problems. The emphasis on mental calculation techniques and shortcut methods reflects the pragmatic nature of Vedic Mathematics, allowing individuals to perform calculations swiftly and accurately, saving time and effort.

Spiritual Growth and Self-Transformation: Vedic Mathematics recognizes the potential for spiritual growth and self-transformation through the study of numbers. It views mathematics as a path to higher consciousness and self-realization. The process of engaging with numbers and exploring the underlying patterns and relationships is seen as a transformative journey that expands one's awareness and deepens one's connection with the universe.

By understanding and embracing the philosophical foundations of Vedic Mathematics, individuals can transcend the conventional boundaries of mathematics and approach it as a means of self-discovery, intellectual growth, and harmonious interaction with the world around them. Vedic Mathematics offers a unique perspective that integrates the mathematical and the spiritual, allowing individuals to explore the interconnectedness between themselves, numbers, and the cosmos.

In summary, the philosophical foundations of Vedic Mathematics emphasize the unity and interconnectedness of the universe, the pursuit of cosmic harmony and balance, the importance of intuition and insight, the practicality and applicability of mathematical techniques, and the potential for spiritual growth and self-transformation. Understanding these foundations enriches our engagement with Vedic Mathematics, providing a deeper understanding of its purpose, methods, and impact on our mathematical thinking and spiritual journey.

C. The connection between Vedic Mathematics and Vedic literature:

The Vedic period in ancient India, dating back to approximately 1500 BCE to 500 BCE, gave rise to not only the Vedic texts but also the development of Vedic Mathematics. The connection between Vedic Mathematics and Vedic literature runs deep, as both are rooted in the profound wisdom and knowledge encapsulated in the Vedas and related texts.

Source of Mathematical Insights: The Vedas, which include the Rigveda, Samaveda, Yajurveda, and Atharvaveda, form the foundational texts of Indian philosophy, spirituality, and knowledge. Within these sacred texts, there are references to mathematical concepts, numerical symbolism, and calculations. Vedic Mathematics draws inspiration from these mathematical insights and symbolic representations embedded in the Vedic literature.

Foundational Principles: Vedic Mathematics shares its foundational principles with Vedic literature. The philosophical underpinnings, such as the recognition of the interconnectedness of all things and the understanding of the cosmos as a manifestation of consciousness, are reflected in both Vedic Mathematics and the Vedic texts. The Vedic concepts of unity, harmony, balance, and cosmic order form the bedrock for the development of Vedic Mathematics.

Sutras and Formulations: The Vedic literature serves as a source for the concise and memorable statements called Sutras in Vedic Mathematics. These Sutras encapsulate mathematical techniques and principles. They provide a framework for problem-solving, often using poetic and mnemonic formulations to aid in memorization and transmission of knowledge. The Sutras in Vedic Mathematics derive their inspiration from the poetic and rhythmic style found in the Vedic literature.

Symbolism and Metaphysics: Vedic Mathematics utilizes symbolism and metaphysics to convey mathematical concepts and principles. The Vedic literature employs symbolic representations to convey deeper meanings and cosmic truths. Similarly, Vedic Mathematics employs symbolic representations, such as the use of dots, lines, and patterns, to express mathematical relationships and calculations. This shared use of symbolism creates a connection between the language of mathematics and the symbolic language of the Vedic literature.

Spiritual and Practical Integration: Vedic Mathematics bridges the gap between the spiritual and practical aspects of mathematics, aligning with the holistic approach found in Vedic literature. Vedic texts recognize the profound spiritual significance of numbers and mathematical concepts. Vedic Mathematics incorporates this spiritual dimension while also providing practical techniques for mental calculation and problem-solving. The integration of the spiritual and practical aspects reflects the convergence of Vedic Mathematics with the overarching philosophy of the Vedic literature.

In summary, Vedic Mathematics and Vedic literature are intertwined through their shared philosophical foundations, the incorporation of mathematical insights from Vedic texts, the utilization of poetic and mnemonic formulations, the use of symbolism and metaphysics, and the integration of spiritual and practical dimensions. The connection between Vedic Mathematics and Vedic literature allows individuals to explore mathematics not only as a logical and analytical discipline but also as a pathway to understanding the deeper truths of the universe, as illuminated by the Vedic texts.

Chapter III
The Sutras: Fundamental Principles of Vedic Mathematics

The Sutras, which are the fundamental principles of Vedic Mathematics: The Sutras in Vedic Mathematics are a set of concise and memorable statements that serve as fundamental principles or guiding rules for solving mathematical problems. These Sutras encapsulate the essence of Vedic Mathematics and provide a framework for performing calculations efficiently and effectively. Each Sutra represents a specific mathematical operation or concept and offers a unique approach to problem-solving. Here, we will explore some of the key Sutras and their significance:

Ekadhikena Purvena (By One More Than the Previous): This Sutra is used for quick mental calculations of squaring numbers ending in 5. According to the Sutra, when squaring a number ending in 5, multiply the preceding digit by itself plus one and append 25 at the end. For example, to square 35, apply the Sutra as follows: (3 x (3 + 1)) 25 = 1225.

Nikhilam Navatashcaramam Dashatah (All From 9 and the Last From 10): This Sutra is useful for multiplication by complements of 10. It involves finding the difference between a number and its nearest multiple of 10, then multiplying it by the complement of that difference, and subtracting the result from the nearest multiple of 10. This method simplifies calculations and reduces the number of steps required. For example, to multiply 98 by 96, we can apply the Sutra as follows: (100 - 2) x (100 - 4) = 9408.

Urdhva-Tiryagbhyam (Vertically and Crosswise): This Sutra is applicable for multiplication of numbers that have complementary differences. It involves multiplying the sum of the differences between the numbers and a reference point (usually a power of 10) by the reference point, and adding the cross-products of the differences. This technique allows for faster and more efficient multiplication. For example, to multiply 13 by 17, apply the Sutra as follows: $(10 + 3) \times (10 + 7) + (3 \times 7) = 221$.

Anurupyena (Proportionality): This Sutra is used for proportional division. It simplifies the process of dividing numbers that are in a ratio. According to the Sutra, divide the first number by the common factor, and multiply the result by the common factor of the second number. This technique ensures that the division remains in proportion. For example, to divide 64 by 4/7, apply the Sutra as follows: $(64 \div 4) \times 7 = 112$.

These are just a few examples of the Sutras used in Vedic Mathematics. The beauty of these principles lies in their simplicity, efficiency, and adaptability to various mathematical operations. The Sutras provide alternative approaches to conventional methods, promoting mental calculation, speed, and accuracy. They also emphasize the interconnectedness of mathematical concepts, enabling individuals to explore the relationships and patterns between numbers and operations.

By applying the Sutras, individuals can develop mental agility, enhance problem-solving skills, and gain a deeper understanding of mathematical concepts. The Sutras not only offer practical techniques but also embody the philosophical foundations of Vedic Mathematics, fostering a holistic and intuitive approach to mathematics. Through the mastery of these Sutras, individuals can navigate through mathematical challenges with ease and explore the limitless possibilities of mathematical exploration.

A. Introduction to the 16 Sutras of Vedic Mathematics:

The 16 Sutras of Vedic Mathematics are the core principles that form the foundation of this ancient mathematical system. These Sutras, derived from the ancient Vedic texts, provide concise and powerful techniques for solving mathematical problems. Each Sutra represents a specific mathematical operation or concept and offers a unique approach to problem-solving. Together, they encompass a wide range of mathematical operations and provide efficient methods for mental calculations.

Let's explore the 16 Sutras and their applications:

1. Ekadhikena Purvena (By One More Than the Previous): This Sutra is used for quick mental calculations of squares, particularly for numbers ending in 5. It simplifies squaring by multiplying the preceding digit by itself plus one and appending 25 at the end.

2. Nikhilam Navatashcaramam Dashatah (All From 9 and the Last From 10): This Sutra is useful for multiplication by complements of 10. It involves finding the difference between a number and its nearest multiple of 10, multiplying it by the complement of that difference, and subtracting the result from the nearest multiple of 10.

3. Urdhva-Tiryagbhyam (Vertically and Crosswise): This Sutra is applicable for multiplication of numbers with complementary differences. It involves multiplying the sum of the differences between the numbers and a reference point by the reference point, and adding the cross-products of the differences.

4. Paravartya Yojayet (Transpose and Adjust): This Sutra simplifies division by transposing the divisor and adjusting the quotient accordingly. It helps in dividing numbers efficiently, especially when the divisor is close to a power of 10.

5. Sunyam Samya Samuccaye (When the Sum is the Same, that Sum is Zero): This Sutra is used for solving linear equations involving sums or differences. It states that if the sum of two numbers is the same as the sum of their differences, then the common sum is zero.

6. Anurupyena (Proportionality): This Sutra simplifies proportional division. It involves dividing the first number by the common factor and multiplying the result by the common factor of the second number.

7. Sankalana-Vyavakalanabhyam (By Addition and By Subtraction): This Sutra offers techniques for mental addition and subtraction. It provides shortcuts for performing calculations by adding or subtracting numbers from a base value.

8. Puranapuranabhyam (By the Completion or Non-Completion): This Sutra simplifies multiplication by complementing the numbers involved. It involves completing or complementing the numbers to simplify the multiplication process.

9. Chalana-Kalanabyham (Differences and Similarities): This Sutra is used for algebraic factorization. It involves identifying differences or similarities between terms and applying appropriate operations to simplify the expression.

10. Yaavadunam (Whatever the Extent of its Deficiency): This Sutra is useful for solving equations involving fractions. It involves estimating the deficiency of a fraction from a reference value and using it to determine the value of the unknown.

11. Vyashtisamanstih (Part and Whole): This Sutra simplifies the calculation of percentages and ratios. It involves expressing the part and the whole as fractions or proportions and using them to determine the required value.

12. Shesanyankena Charamena (The Remainder by the Last Digit): This Sutra offers a quick method for finding remainders when dividing by numbers ending in 9, 8, 7, etc. It involves finding the remainder by the last digit of the divisor.

13. Sopaantyadvayamantyam (The Ultimate and Twice the Penultimate): This Sutra simplifies calculations involving quadratic equations. It involves finding the sum and product of the roots of a quadratic equation using the ultimate and penultimate coefficients.

14. Ekanyunena Purvena (By One Less Than the Previous): This Sutra simplifies calculations involving square roots. It involves finding the square root of a number by considering the preceding digit and subtracting one.

15. Shunyam Saamyasamuccaye (The Sum of Zeroes is Zero): This Sutra provides techniques for multiplying numbers by zero or adding zero to a number. It states that any number multiplied by zero is zero, and the sum of zero and any number is that number itself.

16. Anurupye Shunyamanyat (If One is in Ratio, the Other is Zero): This Sutra simplifies calculations involving algebraic equations and ratios. It states that if one quantity is in a specific ratio to another, and the first quantity becomes zero, then the other quantity is also zero.

These 16 Sutras offer a comprehensive set of techniques and methods for solving mathematical problems. They enable mental calculations, simplify complex operations, and provide shortcuts for efficient problem-solving. By mastering these Sutras, individuals can enhance their mathematical skills, improve their speed and accuracy, and develop a deeper understanding of mathematical concepts.

In summary, the 16 Sutras of Vedic Mathematics encompass a wide range of mathematical operations and provide efficient methods for mental calculations. Each Sutra represents a specific principle or technique, offering alternative approaches to conventional methods. By applying these Sutras, individuals can unlock the power of Vedic

Mathematics and experience the simplicity, speed, and elegance of this ancient mathematical system.

1. "Ekadhikena Purvena"(By One More Than the Previous)

"Ekadhikena Purvena" is a Sutra of Vedic Mathematics that is used for quick mental calculations of squares, particularly for numbers ending in 5. It simplifies the squaring process by multiplying the preceding digit by itself plus one and appending 25 at the end.

Let's illustrate the application of Ekadhikena Purvena with a few examples:

Example 1: Square of 35 To find the square of 35 using Ekadhikena Purvena, follow these steps:

Identify the preceding digit, which is 3 in this case.

Multiply the preceding digit by itself plus one: $3 \times (3 + 1) = 3 \times 4 = 12$.

Append 25 at the end: 1225. Hence, the square of 35 is 1225.

Example 2: Square of 65 To find the square of 65 using Ekadhikena Purvena, follow these steps:

Identify the preceding digit, which is 6 in this case.

Multiply the preceding digit by itself plus one: $6 \times (6 + 1) = 6 \times 7 = 42$.

Append 25 at the end: 4225. Hence, the square of 65 is 4225.

Example 3: Square of 95 To find the square of 95 using Ekadhikena Purvena, follow these steps:

Identify the preceding digit, which is 9 in this case.

Multiply the preceding digit by itself plus one: 9 x (9 + 1) = 9 x 10 = 90.

Append 25 at the end: 9025. Hence, the square of 95 is 9025.

As you can see, Ekadhikena Purvena simplifies the process of finding squares by breaking it down into two steps: multiplying the preceding digit by itself plus one and appending 25 at the end. This technique reduces the complexity of squaring numbers ending in 5 and allows for faster
mental calculations.

It's worth noting that this Sutra can be extended to numbers with more than two digits as well. For example, to find the square of 105, we would use the preceding digit (10) and apply the same steps: 10 x (10 + 1) = 10 x 11 = 110, and then append 25 at the end, resulting in 11025 as the square of 105.

Ekadhikena Purvena is a powerful tool for mentally calculating squares and can be particularly useful in scenarios where quick approximations or estimations are required.

2. "Nikhilam Navatashcaramam Dashatah" (All From 9 and the Last From 10)

"Nikhilam Navatashcaramam Dashatah" is a Sutra of Vedic Mathematics that simplifies multiplication by using complements of 10. It involves finding the difference between a number and its nearest multiple of 10, multiplying it by the complement of that difference, and subtracting the result from the nearest multiple of 10.

Let's illustrate the application of Nikhilam Navatashcaramam Dashatah with a few examples:

Example 1: Multiplication of 98 and 96 To multiply 98 by 96 using Nikhilam Navatashcaramam Dashatah, follow these steps:

Identify the nearest multiple of 10 to each number. In this case, it is 100.

Find the difference between each number and its nearest multiple of 10:

For 98: 100 - 98 = 2.

For 96: 100 - 96 = 4.

Multiply the difference by the complement of that difference:

For 98: 2 x (10 - 2) = 2 x 8 = 16.

For 96: 4 x (10 - 4) = 4 x 6 = 24.

Subtract the products from the nearest multiple of 10:

For 98: 100 - 16 = 84.

For 96: 100 - 24 = 76.

Multiply the two resulting numbers: 84 x 76 = 6384. Hence, the product of 98 and 96 is 6384.

Example 2: Multiplication of 106 and 104 To multiply 106 by 104 using Nikhilam Navatashcaramam Dashatah, follow these steps:

Identify the nearest multiple of 10 to each number. In this case, it is 110.

Find the difference between each number and its nearest multiple of 10:

For 106: 110 - 106 = 4.

For 104: 110 - 104 = 6.

Multiply the difference by the complement of that difference:

For 106: 4 x (10 - 4) = 4 x 6 = 24.

For 104: 6 x (10 - 6) = 6 x 4 = 24.

Subtract the products from the nearest multiple of 10:

For 106: 110 - 24 = 86.

For 104: 110 - 24 = 86.

Multiply the two resulting numbers: 86 x 86 = 7396. Hence, the product of 106 and 104 is 7396.

Nikhilam Navatashcaramam Dashatah simplifies multiplication by breaking it down into simpler steps. It takes advantage of the fact that the difference between a number and its nearest multiple of 10 can be easily computed. By using complements of 10, this Sutra enables faster mental calculations and reduces the complexity of multiplication.

It's important to note that this technique can be extended to larger numbers as well. The principles remain the same: find the difference between each number and its nearest multiple of 10, multiply the differences by the complements, and subtract the results from the nearest multiples of 10.

Nikhilam Navatashcaramam Dashatah provides a powerful tool for mental calculations and is particularly useful when dealing with larger numbers or when quick approximations are required.

B3. Urdhva-Tiryagbhyam (Vertically and Crosswise):

"Urdhva-Tiryagbhyam" is a Sutra of Vedic Mathematics that provides a technique for efficient multiplication called "Vertically and Crosswise." It allows for multiplication of numbers with complementary differences by multiplying the sum of the differences and a reference point by the reference point and adding the cross-products of the differences.

Let's illustrate the application of Urdhva-Tiryagbhyam with a few examples:

Example 1: Multiplication of 14 and 16 To multiply 14 by 16 using Urdhva-Tiryagbhyam, follow these steps:

Identify the differences between the numbers and a reference point. In this case, the reference point can be any power of 10. Let's use 10 for simplicity.

Difference for 14: 14 - 10 = 4.

Difference for 16: 16 - 10 = 6.

Multiply the sum of the differences (4 + 6) by the reference point (10): (4 + 6) × 10 = 10 × 10 = 100.

Multiply the differences crosswise: 4 × 6 = 24.

Add the products obtained in steps 2 and 3: 100 + 24 = 124. Hence, the product of 14 and 16 is 124.

Example 2: Multiplication of 23 and 27 To multiply 23 by 27 using Urdhva-Tiryagbhyam, follow these steps:

Identify the differences between the numbers and a reference point (10):

Difference for 23: 23 - 10 = 13.

Difference for 27: 27 - 10 = 17.

Multiply the sum of the differences (13 + 17) by the reference point (10): (13 + 17) × 10 = 30 × 10 = 300.

Multiply the differences crosswise: 13 × 17 = 221.

Add the products obtained in steps 2 and 3: 300 + 221 = 521. Hence, the product of 23 and 27 is 521.

Urdhva-Tiryagbhyam simplifies multiplication by breaking it down into steps that involve the sum of differences and cross-products. It takes advantage of the relationship between the differences and their cross-products to perform efficient multiplication.

This technique can be extended to larger numbers as well. Simply follow the same steps of identifying the differences, calculating the sum of differences, multiplying it by the reference point, and adding the cross-products. The Urdhva-Tiryagbhyam method allows for mental calculations and reduces the complexity of multiplication, especially when dealing with numbers that have complementary differences.

It's worth noting that the reference point can be any convenient number, and the choice of reference point does not affect the final result. The key is to maintain the relationship between the differences and their cross-products.

Urdhva-Tiryagbhyam provides a powerful and efficient method for multiplication, enabling quick mental calculations and reducing the number of steps required for finding products.

4. Paravartya Yojayet (Transpose and Adjust)

"Paravartya Yojayet" is a Sutra of Vedic Mathematics that simplifies division by transposing the divisor and adjusting the quotient accordingly. It helps in dividing numbers efficiently, especially when the divisor is close to a power of 10.

Let's illustrate the application of Paravartya Yojayet with a few examples:

Example 1: Division of 532 by 48 To divide 532 by 48 using Paravartya Yojayet, follow these steps:

Identify the divisor and dividend: Divisor = 48, Dividend = 532.

Transpose the digits of the divisor: Transposed Divisor = 84.

Divide the transposed divisor into the dividend: 532 ÷ 84 = 6 with a remainder of 28.

Adjust the quotient: Since the divisor was transposed, the quotient needs to be adjusted. Subtract the difference between the divisor and the transposed divisor from the quotient: 6 - (84 - 48) = 6 - 36 = 0. Hence, the division of 532 by 48 using Paravartya Yojayet gives a quotient of 60 with a remainder of 0.

Example 2: Division of 764 by 86 To divide 764 by 86 using Paravartya Yojayet, follow these steps:

Identify the divisor and dividend: Divisor = 86, Dividend = 764.

Transpose the digits of the divisor: Transposed Divisor = 68.

Divide the transposed divisor into the dividend: 764 ÷ 68 = 11 with a remainder of 36.

Adjust the quotient: Subtract the difference between the divisor and the transposed divisor from the quotient: 11 - (86 - 68) = 11 - 18 = - 7. Note: A negative quotient indicates an error or inconsistency. In this case, we need to adjust the negative quotient by adding multiples of the divisor until the quotient becomes positive.

Add 86 to the quotient: -7 + 86 = 79. Hence, the division of 764 by 86 using Paravartya Yojayet gives a quotient of 79 with a remainder of 36.

Paravartya Yojayet simplifies division by transposing the divisor and adjusting the quotient accordingly. This technique eliminates the need for long division and allows for faster mental calculations, especially when dealing with divisors that are close to a power of 10. It streamlines the division process and provides a more efficient approach.

It's important to note that Paravartya Yojayet is most effective when the transposed divisor remains within a manageable range. If the transposed divisor becomes too large or complex, it may be more practical to use other division methods.

Paravartya Yojayet provides a powerful tool for mental calculations and simplifies the division process, particularly when the divisor is close to a power of 10 or can be easily transposed.

5. Sunyam Samya Samuccaye (When the Sum is the Same, that Sum is Zero)

"Sunyam Samya Samuccaye" is a Sutra of Vedic Mathematics that is used for solving linear equations involving sums or differences. It states that if the sum of two numbers is the same as the sum of their differences, then the common sum is zero.

Let's illustrate the application of Sunyam Samya Samuccaye with a few examples:

Example 1: Solving the equation $x + y = x - y$ To solve the equation $x + y = x - y$ using Sunyam Samya Samuccaye, follow these steps:

Recognize that the sum of the two numbers $(x + y)$ is equal to the sum of their differences $(x - y)$.

According to the Sutra, when the sum of the numbers is equal to the sum of their differences, the common sum is zero.

Therefore, $x + y = x - y$ simplifies to $2y = 0$.

Solve for y by dividing both sides by 2: $y = 0$.

Substitute the value of y back into the equation to solve for x: $x + 0 = x - 0$, which simplifies to $x = x$.

The equation $x + y = x - y$ is satisfied when $y = 0$ and any value can be assigned to x. For example, if $x = 5$, then the equation becomes $5 + 0 = 5 - 0$, which is true.

Hence, the solution to the equation $x + y = x - y$ is $y = 0$ and x can be any real number.

Example 2: Solving the equation $(a + b) - (c + d) = (a - c) + (b - d)$
To solve the equation $(a + b) - (c + d) = (a - c) + (b - d)$ using Sunyam Samya Samuccaye, follow these steps:

Recognize that the sum of $(a + b)$ and $(c + d)$ is equal to the sum of $(a - c)$ and $(b - d)$.

According to the Sutra, when the sum of two numbers is equal to the sum of their differences, the common sum is zero.

Therefore, $(a + b) - (c + d) = (a - c) + (b - d)$ simplifies to $0 = 0$.

The equation $0 = 0$ is true, indicating that the equation holds for any values of a, b, c, and d.

Hence, the equation $(a + b) - (c + d) = (a - c) + (b - d)$ is always satisfied regardless of the values of a, b, c, and d.

Sunyam Samya Samuccaye allows for the simplification of linear equations involving sums and differences. By recognizing that the sum of the numbers is equal to the sum of their differences, we can deduce that the common sum is zero. This Sutra provides a useful tool for solving equations and identifying relationships between variables.

It's important to note that while Sunyam Samya Samuccaye is applicable to linear equations, it may not be applicable or useful for equations involving higher degrees or more complex algebraic expressions.

6. Anurupyena (Proportionality)

"Anurupyena" is a Sutra of Vedic Mathematics that simplifies proportional division. It involves dividing the first number by a common factor and multiplying the result by the common factor of the second number. This technique ensures that the division remains in proportion.

Let's illustrate the application of Anurupyena with a few examples:

Example 1: Division of 64 by 4/7 To divide 64 by 4/7 using Anurupyena, follow these steps:

Identify the first number, which is 64, and the second number, which is 4/7.

Identify the common factor between the two numbers, which is 4.

Divide the first number (64) by the common factor (4): $64 \div 4 = 16$.

Multiply the result (16) by the common factor of the second number (7): $16 \times 7 = 112$. Hence, dividing 64 by 4/7 using Anurupyena gives a result of 112.

Example 2: Division of 120 by 2/5 To divide 120 by 2/5 using Anurupyena, follow these steps:

Identify the first number, which is 120, and the second number, which is 2/5.

Identify the common factor between the two numbers, which is 2.

Divide the first number (120) by the common factor (2): 120 ÷ 2 = 60.

Multiply the result (60) by the common factor of the second number (5): 60 × 5 = 300. Hence, dividing 120 by 2/5 using Anurupyena gives a result of 300.

Anurupyena simplifies division by ensuring proportionality between the numbers involved. It involves dividing the first number by the common factor and multiplying the result by the common factor of the second number. This technique allows for quick and efficient division while maintaining the proportional relationship between the numbers.

It's important to note that Anurupyena is most effective when dealing with fractions or ratios that are in proportion to each other. It simplifies calculations by reducing the numbers to manageable forms without altering their proportional relationship.

Anurupyena provides a powerful tool for mental calculations and simplifies division by maintaining proportionality. It allows for efficient computation and ensures accurate results, particularly when dealing with ratios or fractions.

7. Sankalana-Vyavakalanabhyam (By Addition and By Subtraction)

"Sankalana-Vyavakalanabhyam" is a Sutra of Vedic Mathematics that provides techniques for mental addition and subtraction. It offers shortcuts for performing calculations by adding or subtracting numbers from a base value. This Sutra breaks down calculations into smaller additions or subtractions, reducing complexity and cognitive load.

Let's elaborate on the concept of Sankalana-Vyavakalanabhyam and provide examples to illustrate its application:

The Sutra "Sankalana-Vyavakalanabhyam" can be understood as follows:

Addition (Sankalana): When adding numbers, you can mentally perform additions by adding or subtracting the excess or shortfall from a base number. This is done by considering the difference between the numbers and adding or subtracting it from a reference point.

Subtraction (Vyavakalana): Similarly, when subtracting numbers, you can mentally perform subtractions by adding or subtracting the excess or shortfall from a base number. Again, this is done by considering the difference between the numbers and adding or subtracting it from a reference point.

Let's illustrate the application of Sankalana-Vyavakalanabhyam with a few examples:

Example 1: Addition using Sankalana To mentally perform the addition 87 + 38 using Sankalana-Vyavakalanabhyam, follow these steps:

Choose a base number. Let's use 100 as the base.

Determine the excess or shortfall of each number from the base:

For 87: Excess = 87 - 100 = -13.

For 38: Shortfall = 100 - 38 = 62.

Add or subtract the excess or shortfall from the base:

For 87: 100 - 13 = 87.

For 38: 100 + 62 = 162.

Add the adjusted values: 87 + 162 = 249. Hence, the sum of 87 and 38 using Sankalana-Vyavakalanabhyam is 249.

Example 2: Subtraction using Vyavakalana To mentally perform the subtraction 125 - 79 using Sankalana-Vyavakalanabhyam, follow these steps:

Choose a base number. Let's use 100 as the base.

Determine the excess or shortfall of each number from the base:

For 125: Excess = 125 - 100 = 25.

For 79: Shortfall = 100 - 79 = 21.

Add or subtract the excess or shortfall from the base:

For 125: 100 + 25 = 125.

For 79: 100 - 21 = 79.

Perform the adjusted subtraction: 125 - 79 = 46. Hence, the result of 125 minus 79 using Sankalana-Vyavakalanabhyam is 46.

Sankalana-Vyavakalanabhyam allows for quick mental calculations by breaking down additions or subtractions into smaller steps. It reduces the complexity of calculations by considering the differences between numbers and adjusting them based on a reference point.

It's important to note that the choice of the base number is flexible. You can choose a base that is convenient for mental calculations. Additionally, Sankalana-Vyavakalanabhyam can be extended to perform calculations with larger numbers or multiple-digit numbers, following the same principles of considering differences and adjusting from a reference point.

Sankalana-Vyavakalanabhyam provides a powerful tool for mental calculations and simplifies addition and subtraction by breaking them down into manageable steps. It enhances mental arithmetic skills and aids in faster, more efficient calculations.

8. Puranapuranabhyam (By the Completion or Non-Completion)

"Puranapuranabhyam" is a Sutra of Vedic Mathematics that provides techniques for mental addition and subtraction by completing or non-completing the numbers involved. It simplifies calculations by breaking them down into smaller steps, reducing complexity and facilitating mental computation.

Let's elaborate on the concept of Puranapuranabhyam and provide examples to illustrate its application:

The Sutra "Puranapuranabhyam" can be understood as follows:

Completion (Purana): When adding two numbers, A and B, mentally using Puranapuranabhyam:

Identify the difference between each number and a base number, often a power of 10.

Complete the numbers by adding or subtracting the differences to make them easier to work with.

Add the completed numbers to get the final result.

Non-Completion (Aporana): When subtracting one number, B, from another number, A, mentally using Puranapuranabhyam:

Identify the difference between each number and a base number, often a power of 10.

Non-complete the numbers by adding or subtracting the differences to make them easier to work with.

Subtract the non-completed numbers to get the final result.

Let's illustrate the application of Puranapuranabhyam with a few examples:

Example 1: Addition using Puranapuranabhyam To add 97 and 46 using Puranapuranabhyam, follow these steps:

Choose a convenient base number. Let's use 100 as the base.

Determine the difference between each number and the base number:

Difference for 97: 100 - 97 = 3.

Difference for 46: 100 - 46 = 54.

Complete the numbers by adding the differences:

Completed 97: 97 + 3 = 100.

Completed 46: 46 + 54 = 100.

Add the completed numbers: 100 + 100 = 200. Hence, the sum of 97 and 46 using Puranapuranabhyam is 200.

Example 2: Subtraction using Puranapuranabhyam To subtract 57 from 93 using Puranapuranabhyam, follow these steps:

Choose a convenient base number. Let's use 100 as the base.

Determine the difference between each number and the base number:

Difference for 57: 57 - 100 = -43 (adjust the negative difference).

Difference for 93: 93 - 100 = -7 (adjust the negative difference).

Non-complete the numbers by subtracting the differences:

Non-completed 57: 57 - (-43) = 57 + 43 = 100.

Non-completed 93: 93 - (-7) = 93 + 7 = 100.

Subtract the non-completed numbers: 100 - 100 = 0. Hence, the result of subtracting 57 from 93 using Puranapuranabhyam is 0.

Puranapuranabhyam simplifies addition and subtraction by completing or non-completing the numbers involved. By utilizing the differences between each number and a base number, the calculations are broken down into smaller steps, reducing complexity and facilitating mental computation.

It's important to note that the choice of base number should be convenient and close to the numbers being operated on. Using powers of 10 is often helpful, but other base numbers can be used as well.

Puranapuranabhyam provides a powerful tool for mental calculations and simplifies addition and subtraction by completing or non-completing the numbers. It allows for quick and efficient computations, especially when dealing with larger numbers or when mental calculations are required.

9. Chalana-Kalanabyham (Differences and Similarities)

"Chalana-Kalanabyham" is a Sutra of Vedic Mathematics that provides techniques for mental multiplication and division by working with differences and similarities between numbers. It simplifies calculations by breaking them down into smaller steps, reducing complexity and facilitating mental computation.

Let's elaborate on the concept of Chalana-Kalanabyham and provide examples to illustrate its application:

The Sutra "Chalana-Kalanabyham" can be understood as follows:

Differences (Chalana): When multiplying two numbers, A and B, mentally using Chalana-Kalanabyham:

Identify the difference between each number and a base number, often a power of 10.

Compute the difference between the base number and each number.

Calculate the product of these differences.

Adjust the product based on the differences between the base number and the original numbers.

Similarities (Kalanabhyam): When dividing one number, B, by another number, A, mentally using Chalana-Kalanabyham:

Identify the similarity between each number and a base number, often a power of 10.

Compute the difference between the base number and each number.

Calculate the quotient of these differences.

Adjust the quotient based on the differences between the base number and the original numbers.

Let's illustrate the application of Chalana-Kalanabyham with a few examples:

Example 1: Multiplication using Chalana-Kalanabyham To multiply 43 and 52 using Chalana-Kalanabyham, follow these steps:

Choose a convenient base number. Let's use 50 as the base.

Determine the difference between each number and the base number:

Difference for 43: 50 - 43 = 7.

Difference for 52: 50 - 52 = -2 (adjust the negative difference).

Compute the difference between the base number and each number:

Difference for 43: 43 - 50 = -7 (adjust the negative difference).

Difference for 52: 52 - 50 = 2.

Calculate the product of these differences: 7 × 2 = 14.

Adjust the product based on the differences between the base number and the original numbers:

Adjusted product: 50^2 - 14 = 2500 - 14 = 2486. Hence, the product of 43 and 52 using Chalana-Kalanabyham is 2486.

Example 2: Division using Chalana-Kalanabyham To divide 1375 by 25 using Chalana-Kalanabyham, follow these steps:

Choose a convenient base number. Let's use 100 as the base.

Determine the difference between each number and the base number:

Difference for 1375: 1375 - 100 = 1275.

Difference for 25: 25 - 100 = -75 (adjust the negative difference).

Compute the difference between the base number and each number:

Difference for 1375: 100 - 1375 = -1275 (adjust the negative difference).

Difference for 25: 100 - 25 = 75.

Calculate the quotient of these differences: -1275 ÷ 75 = -17.

Adjust the quotient based on the differences between the base number and the original numbers:

Adjusted quotient: 100 - (-17) = 117. Hence, the quotient of 1375 divided by 25 using Chalana-Kalanabyham is 117.

Chalana-Kalanabyham simplifies multiplication and division by working with differences and similarities between numbers. By utilizing the differences between each number and a base number, the calculations are broken down into smaller steps, reducing complexity and facilitating mental computation.

It's important to note that the choice of base number should be convenient and close to the numbers being operated on. Using powers of 10 is often helpful, but other base numbers can be used as well.

Chalana-Kalanabyham provides a powerful tool for mental calculations and simplifies multiplication and division by working with differences and similarities. It allows for quick and efficient computations, especially when dealing with larger numbers or when mental calculations are required.

10. Yaavadunam (Whatever the Extent of its Deficiency)

"Yaavadunam" is a Sutra of Vedic Mathematics that provides a technique for efficient multiplication when one number is deficient (less than a base) and has a specific deficiency. It simplifies calculations by considering the extent of deficiency and adjusting the product accordingly.

Let's elaborate on the concept of Yaavadunam and provide examples to illustrate its application:

The Sutra "Yaavadunam" can be understood as follows:

Deficiency: Let's consider two numbers, A and B, where B is deficient (less than a base number) and has a specific deficiency.

Extent of Deficiency: Determine the extent of deficiency by subtracting B from the base number.

Product Adjustment: Adjust the product of A and B by multiplying it by the extent of deficiency.

Let's illustrate the application of Yaavadunam with a few examples:

Example 1: Multiplication using Yaavadunam To multiply 13 by 8 using Yaavadunam, follow these steps:

Choose a base number. Let's use 10 as the base.

Determine the deficiency of 8 by subtracting it from the base number: 10 - 8 = 2.

Multiply the non-deficient part (3) of 13 by the base number (10): 3 × 10 = 30.

Adjust the product by multiplying it with the extent of deficiency (2): 30 × 2 = 60. Hence, the product of 13 and 8 using Yaavadunam is 60.

Example 2: Multiplication of larger numbers using Yaavadunam To multiply 147 by 92 using Yaavadunam, follow these steps:

Choose a base number. Let's use 100 as the base.

Determine the deficiency of 92 by subtracting it from the base number: 100 - 92 = 8.

Multiply the non-deficient part (47) of 147 by the base number (100): 47 × 100 = 4700.

Adjust the product by multiplying it with the extent of deficiency (8): 4700 × 8 = 37600. Hence, the product of 147 and 92 using Yaavadunam is 37600.

Yaavadunam simplifies multiplication by considering the deficiency of one number and adjusting the product accordingly. It reduces the complexity of multiplication by breaking it down into smaller steps based on the extent of deficiency.

It's important to note that Yaavadunam is most effective when the deficiency is significant compared to the base number. If the deficiency is small, other multiplication techniques may be more practical.

Yaavadunam provides a powerful tool for mental calculations and simplifies multiplication by considering the extent of deficiency. It allows for efficient multiplication, especially when one number is deficient, resulting in faster and more accurate calculations.

11. Vyashtisamanstih (Part and Whole)

"Vyashtisamanstih" is a Sutra of Vedic Mathematics that provides a technique for multiplication by considering the relationship between parts and the whole. It simplifies calculations by breaking down the numbers into their respective parts and then combining them to obtain the final product.

Let's elaborate on the concept of Vyashtisamanstih and provide examples to illustrate its application:

The Sutra "Vyashtisamanstih" can be understood as follows:

Parts and Whole: Let's consider two numbers, A and B, which can be broken down into parts and a whole.

Part Multiplication: Multiply each part of A with each part of B.

Whole Multiplication: Multiply the whole part of A with the whole part of B.

Combine the Results: Combine the products obtained from the part multiplication and the whole multiplication to obtain the final product.

Let's illustrate the application of Vyashtisamanstih with a few examples:

Example 1: Multiplication using Vyashtisamanstih To multiply 34 by 23 using Vyashtisamanstih, follow these steps:

Break down each number into parts and a whole:

Number 34: Parts = 30 and 4, Whole = 34.

Number 23: Parts = 20 and 3, Whole = 23.

Multiply the parts of each number:

Part multiplication: $(30 \times 20) + (30 \times 3) + (4 \times 20) + (4 \times 3) = 600 + 90 + 80 + 12 = 782$.

Multiply the wholes of each number:

Whole multiplication: $34 \times 23 = 782$.

Combine the results: The final product is 782.

Hence, the product of 34 and 23 using Vyashtisamanstih is 782.

Example 2: Multiplication of larger numbers using Vyashtisamanstih
To multiply 267 by 156 using Vyashtisamanstih, follow these steps:

Break down each number into parts and a whole:

Number 267: Parts = 200, 60, and 7, Whole = 267.

Number 156: Parts = 100, 50, and 6, Whole = 156.

Multiply the parts of each number:

Part multiplication: $(200 \times 100) + (200 \times 50) + (200 \times 6) + (60 \times 100) + (60 \times 50) + (60 \times 6) + (7 \times 100) + (7 \times 50) + (7 \times 6) = 20000 + 10000 + 1200 + 6000 + 3000 + 360 + 700 + 350 + 42 = 40452$.

Multiply the wholes of each number:

Whole multiplication: $267 \times 156 = 41772$.

Combine the results: The final product is $40452 + 41772 = 82224$.

Hence, the product of 267 and 156 using Vyashtisamanstih is 82224.

Vyashtisamanstih simplifies multiplication by breaking down the numbers into parts and a whole. By multiplying the parts of each number and the wholes of each number separately, and then combining the results, the complexity of multiplication is reduced.

It's important to note that Vyashtisamanstih is most effective when the numbers can be conveniently broken down into parts and a whole. If the numbers have complex or irregular patterns, other multiplication techniques may be more practical.

Vyashtisamanstih provides a powerful tool for mental calculations and simplifies multiplication by considering the relationship between parts and the whole. It allows for efficient and accurate multiplication, especially when dealing with larger numbers or when mental calculations are required.

12. Shesanyankena Charamena (The Remainder by the Last Digit)

"Shesanyankena Charamena" is a Sutra of Vedic Mathematics that provides a technique for quickly calculating remainders by considering the last digit of a number. It simplifies calculations by focusing on the pattern of the last digit and using it to determine the remainder.

Let's elaborate on the concept of Shesanyankena Charamena and provide examples to illustrate its application:

The Sutra "Shesanyankena Charamena" can be understood as follows:

Last Digit: Let's consider a number, A, and focus on its last digit.

Remainder Pattern: Observe the pattern in which the last digit repeats or changes as A is multiplied or powered by different numbers.

Determine the Remainder: Based on the pattern, find the remainder when A is divided by a particular number or power of 10.

Let's illustrate the application of Shesanyankena Charamena with a few examples:

Example 1: Finding the remainder using Shesanyankena Charamena To find the remainder when 347 is divided by 10 using Shesanyankena Charamena, follow these steps:

Focus on the last digit of the number, which is 7.

Observe the pattern of the last digit as it is multiplied by different powers of 10: 7, 70, 700, 7000, ...

Notice that the last digit repeats in a cycle of 7, 0, 0, 0, ...

Since the pattern repeats every four powers of 10, we can conclude that the remainder when 347 is divided by 10 is 7.

Hence, the remainder when 347 is divided by 10 using Shesanyankena Charamena is 7.

Example 2: Finding the remainder of a larger number using Shesanyankena Charamena To find the remainder when 2896 is divided by 100 using Shesanyankena Charamena, follow these steps:

Focus on the last two digits of the number, which are 96.

Observe the pattern of the last two digits as they are multiplied by different powers of 10: 96, 9600, 960000, ...
Notice that the last two digits repeat in a cycle of 96, 00, 00, ...

Since the pattern repeats every two powers of 10, we can conclude that the remainder when 2896 is divided by 100 is 96.

Hence, the remainder when 2896 is divided by 100 using Shesanyankena Charamena is 96.

Shesanyankena Charamena simplifies the calculation of remainders by focusing on the pattern of the last digit or digits. By observing how the last digit repeats or changes as the number is multiplied or powered by different numbers, we can determine the remainder without performing the full division.

It's important to note that Shesanyankena Charamena is most effective when the remainder pattern is regular and predictable. If the number has a complex or irregular pattern, other techniques may be more suitable.

Shesanyankena Charamena provides a useful tool for mental calculations and simplifies the determination of remainders by considering the last digit or digits. It allows for quick and efficient calculation of remainders, especially when dealing with larger numbers or when mental calculations are required.

13. Sopaantyadvayamantyam (The Ultimate and Twice the Penultimate)

"Sopaantyadvayamantyam" is a Sutra of Vedic Mathematics that provides a technique for finding the product of the ultimate (last) digit and twice the penultimate (second to last) digit of a number. It simplifies calculations by focusing on these specific digits and utilizing their relationship to obtain the desired product.

Let's elaborate on the concept of Sopaantyadvayamantyam and provide examples to illustrate its application:

The Sutra "Sopaantyadvayamantyam" can be understood as follows:

Ultimate Digit: Let's consider a number, A, and focus on its ultimate (last) digit.

Penultimate Digit: Consider the digit immediately preceding the ultimate digit, known as the penultimate (second to last) digit.

Product Calculation: Multiply the ultimate digit by twice the penultimate digit to obtain the desired product.

Let's illustrate the application of Sopaantyadvayamantyam with a few examples:

Example 1: Finding the product using Sopaantyadvayamantyam To find the product of the ultimate digit and twice the penultimate digit of the number 758, using Sopaantyadvayamantyam, follow these steps:

The ultimate digit of 758 is 8.

The penultimate digit of 758 is 5.

Multiply the ultimate digit (8) by twice the penultimate digit (2 * 5 = 10): 8 * 10 = 80. Hence, the product of the ultimate digit and twice the penultimate digit of 758 using Sopaantyadvayamantyam is 80.

Example 2: Finding the product of a larger number using Sopaantyadvayamantyam To find the product of the ultimate digit and twice the penultimate digit of the number 6,324, using Sopaantyadvayamantyam, follow these steps:

The ultimate digit of 6,324 is 4.

The penultimate digit of 6,324 is 2.

Multiply the ultimate digit (4) by twice the penultimate digit (2 * 2 = 4): 4 * 4 = 16. Hence, the product of the ultimate digit and twice the penultimate digit of 6,324 using Sopaantyadvayamantyam is 16.

Sopaantyadvayamantyam simplifies calculations by focusing on the relationship between the ultimate digit and twice the penultimate digit. By utilizing these specific digits and multiplying them, the desired product can be obtained.

It's important to note that Sopaantyadvayamantyam is most effective when the relationship between the ultimate digit and twice the penultimate digit is applicable to the number being considered. If the number does not follow this pattern, other techniques may be more suitable.

Sopaantyadvayamantyam provides a useful tool for mental calculations and simplifies finding the product of specific digits. It allows for quick and efficient multiplication, especially when dealing with larger numbers or when mental calculations are required.

14. Ekanyunena Purvena (By One Less Than the Previous)

"Ekanyunena Purvena" is a Sutra of Vedic Mathematics that provides a technique for multiplying a number by one less than the previous number. It simplifies calculations by utilizing the relationship between consecutive numbers and finding the product based on that relationship.

Let's elaborate on the concept of Ekanyunena Purvena and provide examples to illustrate its application:
The Sutra "Ekanyunena Purvena" can be understood as follows:

Previous Number: Let's consider a number, A, and focus on its previous number, which is A - 1.

Product Calculation: Multiply A by (A - 1) to obtain the desired product.

Let's illustrate the application of Ekanyunena Purvena with a few examples:

Example 1: Finding the product using Ekanyunena Purvena To find the product of 9 and one less than 9 using Ekanyunena Purvena, follow these steps:

The previous number of 9 is 9 - 1 = 8.

Multiply 9 by 8: 9 * 8 = 72. Hence, the product of 9 and one less than 9 using Ekanyunena Purvena is 72.

Example 2: Finding the product of a larger number using Ekanyunena Purvena To find the product of 20 and one less than 20 using Ekanyunena Purvena, follow these steps:

The previous number of 20 is 20 - 1 = 19.

Multiply 20 by 19: 20 * 19 = 380. Hence, the product of 20 and one less than 20 using Ekanyunena Purvena is 380.

Ekanyunena Purvena simplifies calculations by utilizing the relationship between a number and its previous number. By recognizing that the desired product can be obtained by multiplying the number by one less than itself, the calculation becomes more straightforward.

It's important to note that Ekanyunena Purvena is most effective when the relationship between the number and its previous number is applicable to the context in which it is used. If the pattern does not align, other techniques may be more suitable.

Ekanyunena Purvena provides a useful tool for mental calculations and simplifies multiplication by leveraging the relationship between consecutive numbers. It allows for quick and efficient multiplication, especially when dealing with consecutive or closely related numbers.

15. Shunyam Saamyasamuccaye (The Sum of Zeroes is Zero)

"Shunyam Saamyasamuccaye" is a Sutra of Vedic Mathematics that states that the sum of any number of zeroes is always zero. It simplifies calculations by recognizing the pattern and property of zero, where adding zero to any number does not change its value.

Let's elaborate on the concept of Shunyam Saamyasamuccaye and provide examples to illustrate its application:
The Sutra "Shunyam Saamyasamuccaye" can be understood as follows:

Zeroes: Consider any number of zeroes (0) to be added together.

Sum Calculation: Regardless of the quantity of zeroes, the sum will always be zero.

Let's illustrate the application of Shunyam Saamyasamuccaye with a few examples:

Example 1: Adding zeroes using Shunyam Saamyasamuccaye To add two zeroes using Shunyam Saamyasamuccaye, follow these steps:

Consider the zeroes to be added: 0 + 0.

Apply the principle of Shunyam Saamyasamuccaye: 0 + 0 = 0. Hence, the sum of two zeroes using Shunyam Saamyasamuccaye is 0.

Example 2: Adding multiple zeroes using Shunyam Saamyasamuccaye To add five zeroes using Shunyam Saamyasamuccaye, follow these steps:

Consider the zeroes to be added: $0 + 0 + 0 + 0 + 0$.

Apply the principle of Shunyam Saamyasamuccaye: $0 + 0 + 0 + 0 + 0 = 0$. Hence, the sum of five zeroes using Shunyam Saamyasamuccaye is 0.

Shunyam Saamyasamuccaye simplifies calculations by recognizing the inherent property of zero. Zero is known as an additive identity, meaning that adding zero to any number does not change its value. Therefore, regardless of the quantity of zeroes being added, the sum will always be zero.

It's important to note that the application of Shunyam Saamyasamuccaye is specific to zeroes. Adding any other number along with zeroes will yield a different result. For example, adding $5 + 0$ would result in 5, as zero does not affect the value of the non-zero number.

Shunyam Saamyasamuccaye provides a powerful tool for mental calculations and simplifies addition by recognizing the pattern of zero. It allows for quick and efficient calculations when working with zeroes, ensuring that the sum is always zero.

16. Anurupye Shunyamanyat (If One is in Ratio, the Other is Zero)

"Anurupye Shunyamanyat" is a Sutra of Vedic Mathematics that states that if one number is in a certain ratio with another number, and the first number is zero, then the other number is also zero. It simplifies calculations by recognizing the property of zero in proportionality.

Let's elaborate on the concept of Anurupye Shunyamanyat and provide examples to illustrate its application:

The Sutra "Anurupye Shunyamanyat" can be understood as follows:

Proportionality: Consider two numbers, A and B, where A is in a specific ratio with B.

Condition: If A is zero, then B will also be zero.

Let's illustrate the application of Anurupye Shunyamanyat with a few examples:

Example 1: Proportionality with zero using Anurupye Shunyamanyat If the ratio of A to B is 0:7, and A is zero, then according to Anurupye Shunyamanyat, B will also be zero. This means that if one number in a ratio is zero, the other number in that ratio will also be zero.

Example 2: Proportionality with zero in a larger context using Anurupye Shunyamanyat If the ratio of A to B is 0:5, and A is zero, then Anurupye Shunyamanyat states that B will also be zero. Similarly, if the ratio of C to D is 0:3, and C is zero, then D will also be zero. This pattern holds true for any ratio where one number is zero.

Anurupye Shunyamanyat simplifies calculations by recognizing the property of zero in proportionality. If one number in a ratio is zero, then the other number will also be zero, regardless of the specific ratio.

It's important to note that the application of Anurupye Shunyamanyat is specific to the context of proportionality and ratios. It does not apply in other mathematical operations or scenarios where zero may have different implications.

Anurupye Shunyamanyat provides a useful tool for mental calculations and simplifies calculations involving ratios. It allows for quick identification of zero in proportionality, providing a direct relationship between the two numbers.

Chapter IV
Arithmetic Operations

A. Enhancing Mental Calculation Through Vedic Mathematics

Enhancing mental calculation through Vedic Mathematics involves practicing and applying various techniques to perform calculations mentally. Here are some steps to enhance mental calculation skills using Vedic Mathematics:

Familiarize Yourself with Vedic Mathematics Techniques: Start by learning and understanding the different techniques of Vedic Mathematics. Study the sutras (mathematical aphorisms) and their applications. Get acquainted with the principles, formulas, and patterns involved in mental calculations.

Break Down Numbers: Break down numbers into their constituent parts to simplify calculations. For example, represent a number as the sum of its digits or consider its complement to the nearest multiple of 10.

Practice Mental Calculation Drills: Regularly engage in mental calculation drills to improve speed and accuracy. Start with simple calculations and gradually increase the complexity. Focus on performing calculations mentally without relying on pen and paper or calculators.

Visualize the Calculation: Develop the ability to visualize numbers and their relationships. Visualize patterns, digits, and operations in your mind's eye to perform calculations quickly. This skill becomes more intuitive with practice.

Utilize Crosschecking Techniques: Vedic Mathematics provides crosschecking techniques that help verify the correctness of calculations mentally. By applying these techniques, you can identify errors and rectify them efficiently.

Apply Techniques in Real-Life Scenarios: Look for opportunities to apply Vedic Mathematics techniques in real-life scenarios. For instance, use mental calculation methods while shopping, calculating bills, estimating quantities, or solving everyday math problems. This practical application helps reinforce the techniques and strengthens mental calculation skills.

Practice Regularly: Consistent practice is crucial for enhancing mental calculation skills. Set aside dedicated time for mental calculations and incorporate them into your daily routine. Regular practice helps build speed, accuracy, and confidence in mental calculations.

Challenge Yourself: As you progress, challenge yourself with more complex calculations and strive to improve your speed and accuracy. Work on increasing the range of numbers you can handle mentally and aim to perform calculations in less time.

Seek Additional Resources: Explore books, online resources, and tutorials on Vedic Mathematics to deepen your understanding and discover new techniques. There are various Vedic Mathematics resources available that provide step-by-step guidance and practice exercises.

Remember, enhancing mental calculation skills through Vedic Mathematics requires patience, practice, and perseverance. Start with basic calculations and gradually work your way up to more advanced techniques. Over time, you'll notice a significant improvement in your mental calculation abilities, leading to faster and more accurate calculations.

B. Addition and Subtraction Techniques Using Vedic Methods

Let's explore addition and subtraction techniques using Vedic methods with examples:

Addition Techniques:

a. Nikhilam Sutra (All From 9 and the Last From 10):

Example: Add 97 and 96.
Step 1: Complement both numbers to the nearest multiple of 10. In this case, complement 97 to 100 and 96 to 100.
Step 2: Add the complemented numbers: 100 + 100 = 200. Step 3: Subtract the difference from 10: 10 - 0 = 10. The sum of 97 and 96 using the Nikhilam Sutra is 200 - 10 = 190.

b. Ekadhikena Purvena (By One More Than the Previous):

Example: Add 8 and 9.
Step 1: Multiply 9 by one more than the previous number (8 + 1 = 9).
Step 2: 9 × 9 = 81. The sum of 8 and 9 using Ekadhikena Purvena is 81.

Subtraction Techniques:

a. Anurupye Shunyamanyat (If One is in Ratio, the Other is Zero):

Example: Subtract 630 from 630. Since both numbers are the same, the result is zero.

b. Urdhva-Tiryagbhyam (Vertically and Crosswise):

Example: Subtract 547 from 1000.
Step 1: Transpose 547 to 753 and add the numbers: 1000 + 753 = 1753.
Step 2: Adjust the result by subtracting the original 547: 1753 - 547 = 1206. The difference between 1000 and 547 using Urdhva-Tiryagbhyam is 1206.
These examples demonstrate the application of Vedic techniques in performing addition and subtraction calculations. By utilizing these methods, you can simplify complex calculations, improve mental

computation skills, and perform calculations more efficiently. With practice, you'll be able to apply these techniques to various addition and subtraction problems and achieve faster and accurate results.

C. Multiplication Techniques: One-Line and Specific Digit Methods

Vedic Mathematics offers multiplication techniques that can simplify calculations and make them more efficient. Let's explore two popular multiplication techniques: the one-line method and the specific digit method.

One-Line Multiplication Method:

The one-line multiplication method allows you to perform multiplications mentally in a single line, without the need for intermediate steps. It involves breaking down the multiplication into subproblems and combining the results.

Example: Multiply 12 by 15 using the one-line method.

Step 1: Split the numbers into their components: 12 = 10 + 2 and 15 = 10 + 5.

```
   10 + 2
 × 10 + 5
------------------
```

Step 2: Multiply the components diagonally and write the results.

```
   10 + 2
 × 10 + 5
------------------
100 + 50 +
20 + 10
```

Step 3: Add the diagonal results to get the final product.

```
   10 + 2
× 10 + 5
------------------
100 + 50 +
20 + 10
------------------
180 + 20 + 0
```

The product of 12 and 15 using the one-line method is 180 + 20 + 0 = 200.

Specific Digit Multiplication Method:

The specific digit method is useful when multiplying a number by a specific digit, such as multiplying by 9, 99, 999, etc. It involves simple adjustments and additions based on the specific digit.

Example: Multiply 35 by 99 using the specific digit method.

Step 1: Subtract the original number from the specific digit repeated one more time.

```
  99
- 35
---------
  64
```

Step 2: Append the result of Step 1 to the original number.

```
  35
+ 64
---------
 3496
```

The product of 35 and 99 using the specific digit method is 3496.

These multiplication techniques from Vedic Mathematics offer efficient ways to perform multiplications mentally. They simplify complex calculations and provide quicker results. By practicing these methods, you can enhance your mental calculation skills and perform multiplications with greater speed and accuracy.

D. Division Techniques: Duplex Method and Nikhilam Method

Vedic Mathematics offers division techniques that can simplify calculations and make them more efficient. Let's explore two popular division techniques: the duplex method and the Nikhilam method.

Duplex Method:

The duplex method is a division technique that involves finding quotients quickly through a series of steps involving duplexes (two-digit numbers). It simplifies the division process and allows for faster mental calculations.

Example: Divide 1875 by 25 using the duplex method.

Step 1: Identify a duplex close to or slightly higher than the divisor (25). In this case, we can choose 26, which is the next duplex after 25.

Step 2: Divide the dividend (1875) by the duplex (26). This gives the first part of the quotient, which is the whole number part.

26 | 1875

Quotient: 7 (26 fits into 1875 seven times).

Step 3: Multiply the quotient (7) by the duplex (26) to get the product.

7
× 26

Product: 182

Step 4: Subtract the product (182) from the dividend (1875) to get the remainder.

```
  182
- 175
--------
  125
```

Remainder: 125

Step 5: Bring down the next digit of the dividend (5) and place it next to the remainder.

```
  182
- 175
--------
1255
```

Step 6: Repeat steps 2 to 5 with the new dividend (1255) until the dividend is fully divided or until the desired level of accuracy is reached.

The quotient obtained in each step is combined to get the final quotient. In this case, the quotient is 74.

Therefore, 1875 divided by 25 using the duplex method gives a quotient of 74.

Nikhilam Method:

The Nikhilam method is a division technique used for division by a specific digit, such as 9, 99, 999, etc. It simplifies the division process by reducing the number of steps required.

Example: Divide 936 by 9 using the Nikhilam method.

Step 1: Determine the complement of the divisor (9) with respect to the nearest power of 10, which is 10. The complement is obtained by subtracting the digit from 9.

9 - 9 = 0

Step 2: Multiply the dividend (936) by the complement (0) and write the result.

936 × 0 = 0

Therefore, 936 divided by 9 using the Nikhilam method gives a quotient of 0.

These division techniques from Vedic Mathematics offer efficient ways to perform divisions mentally. They simplify complex calculations and provide quicker results. By practicing these methods, you can enhance your mental calculation skills and perform divisions with greater speed and accuracy.

Chapter V
Practical Applications and Relevance

A. Real-World Examples Showcasing the Practicality of Vedic Mathematics

Vedic Mathematics, with its mental calculation techniques, pattern recognition, and simplification methods, can be applied to various real-world scenarios to enhance calculation speed, accuracy, and problem-solving abilities. Here are some examples showcasing the practicality of Vedic Mathematics:

Quick Mental Calculations: Vedic Mathematics techniques allow for faster mental calculations, enabling individuals to perform arithmetic operations swiftly and accurately. This can be beneficial in everyday situations, such as calculating bills, estimating expenses, and making quick decisions while shopping.

Competitive Examinations: Vedic Mathematics is often employed by students preparing for competitive examinations, where time management is crucial. The mental calculation techniques can help solve mathematical problems efficiently, save time during exams, and improve overall performance.

Financial Calculations: Vedic Mathematics principles, such as proportionality and pattern recognition, can aid in financial calculations. For instance, quick estimation techniques can be used for interest calculations, budgeting, investment analysis, and loan calculations.

Business and Accounting: Vedic Mathematics techniques can enhance speed and accuracy in business and accounting tasks. Quick mental calculations facilitate rapid calculations of profit margins, percentages, discounts, and inventory management, enabling better decision-making in various business operations.

Data Analysis and Statistics: Vedic Mathematics techniques can be employed to perform rapid calculations and estimations in data analysis and statistical applications. They help in quickly calculating mean, median, mode, standard deviation, and other statistical measures, enabling efficient data analysis and interpretation.

Geometry and Architecture: Vedic Mathematics promotes pattern recognition and symmetry, which can aid in geometric constructions, architectural designs, and spatial analysis. It can help in making quick measurements, determining proportions, and visualizing complex shapes or structures.

Cryptography and Encryption: As mentioned earlier, certain aspects of Vedic Mathematics, such as prime number generation, modular arithmetic, and mental calculations, can find applications in cryptography and encryption techniques, improving efficiency in key generation and encryption/decryption processes. While Vedic Mathematics may not provide a comprehensive solution to all mathematical problems in the real world, its principles and techniques can be applied to various practical scenarios to enhance calculation skills, save time, and improve problem-solving abilities. It is important to note that the selection and application of Vedic Mathematics techniques should be done judiciously, considering the specific requirements and context of the problem at hand.

B. Enhancing Speed and Accuracy in Competitive Exams

To enhance speed and accuracy in competitive exams using Vedic Mathematics techniques, you can follow these strategies:

Learn Vedic Mathematics Techniques: Familiarize yourself with various Vedic Mathematics techniques, such as sutras (formulae) and mental calculation methods. These techniques can help you perform calculations faster and simplify complex problems.

Practice Mental Calculation: Regularly practice mental calculations using Vedic Mathematics techniques. Start with simple calculations and gradually increase the complexity. Focus on speed and accuracy while performing calculations mentally.

Memorize Key Formulas: Memorize important formulas and patterns that frequently appear in competitive exams. This will allow you to quickly recall and apply the formulas during the exam, saving valuable time.

Develop Number Sense: Work on improving your number sense by developing an intuitive understanding of numbers and their relationships. This will help you make estimations, identify patterns, and quickly analyze numerical data during exams.

Enhance Problem-Solving Skills: Vedic Mathematics emphasizes pattern recognition and problem-solving. Practice solving a wide range of mathematical problems using Vedic techniques, focusing on finding efficient and quick solutions.

Time Management: Time management is crucial in competitive exams. Practice solving questions within time constraints. Use Vedic Mathematics techniques to optimize your speed while maintaining accuracy.

Simultaneous Operations: Vedic Mathematics techniques allow you to perform multiple calculations simultaneously. Practice breaking down complex problems into smaller steps and perform calculations concurrently to save time.

Mock Tests and Previous Papers: Solve mock tests and previous years' question papers to familiarize yourself with the exam format and identify areas where you can apply Vedic Mathematics techniques effectively. Analyze your performance, identify weaknesses, and work on improving them.

Develop a Problem-Specific Approach: Different Vedic Mathematics techniques are suitable for different types of problems. Develop a problem-specific approach, identifying the most appropriate technique to solve each type of question quickly and accurately.

Stay Calm and Focused: Maintaining a calm and focused mindset during exams is crucial. Practice relaxation techniques and develop strategies to stay composed and focused on the task at hand.

Remember, consistent practice is key to improving your speed and accuracy using Vedic Mathematics techniques. Gradually incorporate these techniques into your study routine, and monitor your progress to track your improvement over time.

C. Incorporating Vedic Mathematics Into Daily Life and Professional Fields

Incorporating Vedic Mathematics principles into daily life and professional fields can enhance your mathematical skills, problem-solving abilities, and mental agility. Here are some ways you can apply Vedic Mathematics in various aspects of life:

Everyday Calculations: In daily life, Vedic Mathematics techniques can help in performing quick mental calculations for tasks such as calculating bills, estimating expenses, determining discounts, and managing budgets. It can save time and improve accuracy in routine calculations.

Education and Learning: Students can incorporate Vedic Mathematics techniques into their studies to solve mathematical problems efficiently. It can aid in mental calculation, improve speed, and simplify complex calculations. Vedic Mathematics can be particularly useful for competitive exams, where time management is crucial.

Business and Finance: Vedic Mathematics principles can be applied in business and finance-related tasks. It can assist in financial calculations, analyzing investment returns, assessing profit margins, and managing inventory. The ability to quickly perform calculations mentally can be advantageous in making informed business decisions.

Data Analysis and Statistics: Vedic Mathematics techniques can be used to perform rapid calculations and estimations in data analysis and statistics. It can help in calculating mean, median, mode, standard deviation, and other statistical measures. The speed and mental agility gained from Vedic Mathematics can be valuable in processing and interpreting data.

Architecture and Design: Vedic Mathematics principles, such as symmetry and proportionality, can be applied in architecture and design. It can assist in geometric calculations, spatial analysis, and creating aesthetically pleasing designs. Vedic Mathematics promotes pattern recognition, which can aid in architectural visualization and optimization.

Problem Solving and Decision Making: Vedic Mathematics fosters a problem-solving mindset by emphasizing pattern recognition, logical reasoning, and simplification techniques. It can enhance problem-solving skills in various professional fields, such as engineering, computer science, and management. Vedic Mathematics encourages thinking outside the box and finding creative solutions to complex problems.

Cryptography and Encryption: Certain aspects of Vedic Mathematics, such as prime number generation and modular arithmetic, can find applications in cryptography and encryption techniques. It can aid in key generation, encryption/decryption processes, and security-related calculations.

Remember, incorporating Vedic Mathematics into daily life and professional fields requires consistent practice and application. By continuously honing your mental calculation skills, improving problem-solving abilities, and leveraging Vedic Mathematics principles, you can enhance efficiency, accuracy, and decision-making in various areas of life and work.

Chapter VI
Criticisms and Controversies

A. Addressing Common Criticisms Against Vedic Mathematics

Vedic Mathematics, like any mathematical system or approach, has received its share of criticisms. It is important to address some common criticisms against Vedic Mathematics:

Lack of Rigorous Mathematical Foundation: Critics argue that Vedic Mathematics lacks a formal and rigorous mathematical foundation. While Vedic Mathematics is not a complete mathematical system in itself, it offers mental calculation techniques and shortcuts that can complement traditional mathematical methods.

Limited Applicability: Critics contend that Vedic Mathematics has limited applicability beyond specific calculations. While Vedic Mathematics is primarily focused on mental calculation and simplification techniques, it can be applied to various mathematical concepts and problem-solving scenarios. However, it is not meant to replace or supersede other branches of mathematics.

Historical Accuracy and Origin Claims: Some critics question the historical accuracy and claims of Vedic Mathematics being derived from ancient Vedic texts. The historical connections and origins of Vedic Mathematics are subject to debate, and there are differing views regarding its authenticity and lineage.

Cultural and Religious Bias: Critics argue that Vedic Mathematics carries cultural and religious biases, as it is associated with Vedic traditions and ancient Indian culture. It is important to note that while Vedic Mathematics draws inspiration from ancient texts, its techniques and principles can be studied and applied independently of cultural or religious affiliations.

Lack of Comprehensive Coverage: Vedic Mathematics does not encompass all areas of mathematics. It primarily focuses on mental calculation techniques and algebraic principles. Other branches of mathematics, such as calculus, geometry, and number theory, require dedicated study beyond the scope of Vedic Mathematics.

Standardization and Consistency: Vedic Mathematics lacks a standardized and universally accepted set of techniques and rules. Different sources and teachers may present variations or interpretations of Vedic Mathematics principles. This can lead to confusion and inconsistency when studying or applying Vedic Mathematics.

It is important to approach Vedic Mathematics with an open and critical mindset, considering its techniques as complementary tools rather than a complete mathematical system. Vedic Mathematics can offer alternative approaches, mental calculation strategies, and problem-solving techniques that can supplement traditional mathematical methods. It is recommended to study Vedic Mathematics alongside traditional mathematics, leveraging its strengths where appropriate and acknowledging its limitations.

B. Debunking Misconceptions and Clarifying its Limitations

Misconceptions and limitations of Vedic Mathematics can arise due to misunderstandings or exaggerated claims about its scope and capabilities. Here, let's clarify some common misconceptions and limitations of Vedic Mathematics:

Vedic Mathematics is a Complete Mathematical System:

Misconception: One common misconception is that Vedic Mathematics is a complete mathematical system that encompasses all branches of mathematics.

Clarification: Vedic Mathematics is not a comprehensive mathematical system. It primarily focuses on mental calculation techniques, pattern recognition, and algebraic principles. While it can be useful for specific calculations and problem-solving, it does not replace or encompass all areas of mathematics, such as calculus, geometry, and number theory.

Vedic Mathematics Offers a Shortcut for Complex Calculations:

Misconception: Another misconception is that Vedic Mathematics provides shortcuts or tricks to solve complex mathematical problems effortlessly.

Clarification: Vedic Mathematics does provide mental calculation techniques and shortcuts that can simplify certain calculations. However, these techniques are not applicable to all types of problems and may not always be the most efficient approach. Traditional mathematical methods and techniques remain essential for a comprehensive understanding of mathematical concepts.

Vedic Mathematics is Ancient and Derived from Vedic Texts:

Misconception: There is a misconception that Vedic Mathematics is directly derived from ancient Vedic texts and has an unbroken lineage.

Clarification: The historical origins and authenticity of Vedic Mathematics are subject to debate. While Vedic Mathematics draws inspiration from certain ancient texts, its techniques and principles have been developed and refined by various mathematicians and scholars over time. The direct connection to ancient Vedic texts is not universally accepted by scholars.

Vedic Mathematics is Superior to Traditional Mathematics:

Misconception: It is sometimes claimed that Vedic Mathematics is superior to traditional mathematics and offers a more efficient and intuitive approach to calculations.
Clarification: Vedic Mathematics is not intended to replace or supplant traditional mathematics. It provides alternative techniques and approaches that can be useful in specific situations. Traditional mathematics provides a comprehensive framework and deep understanding of mathematical principles that are necessary for advanced study and application.

Vedic Mathematics is Universally Accepted and Standardized:

Misconception: Some people assume that Vedic Mathematics has a universally accepted set of techniques and rules that are consistent across all sources and practitioners.

Clarification: Vedic Mathematics lacks a standardized and universally accepted set of techniques. Different sources may present variations or interpretations of Vedic Mathematics principles. This can lead to inconsistencies and differences in the application of Vedic Mathematics techniques.

It is important to approach Vedic Mathematics with a balanced perspective, understanding its strengths, limitations, and context. While it can offer valuable mental calculation techniques and problem-solving approaches, it is essential to supplement Vedic Mathematics with a solid foundation in traditional mathematics for a comprehensive understanding of mathematical concepts.

C. Exploring Alternative Viewpoints and Debates

Alternative viewpoints and debates surrounding Vedic Mathematics exist within the academic and mathematical communities. Here are some alternative viewpoints and debates regarding Vedic Mathematics:

Historical Authenticity: One of the debates revolves around the historical authenticity of Vedic Mathematics. Some scholars argue that the techniques presented in Vedic Mathematics have no direct historical connection to ancient Vedic texts. They suggest that Vedic Mathematics is a modern reconstruction or reinterpretation of mathematical concepts with a Vedic framework.

Lack of Formal Mathematical Rigor: Critics argue that Vedic Mathematics lacks the formal rigor found in traditional mathematics. They contend that Vedic Mathematics does not provide a solid foundation for understanding and developing mathematical concepts and theories. The focus on mental calculation techniques and shortcuts may overshadow the need for deeper mathematical understanding.

Scope and Applicability: Debates also arise regarding the scope and applicability of Vedic Mathematics. While proponents of Vedic Mathematics assert its usefulness in mental calculation, critics argue that the techniques have limited applicability beyond specific calculations. They claim that Vedic Mathematics does not provide a comprehensive mathematical framework capable of addressing the wide range of mathematical concepts and problems.

Standardization and Consistency: Another point of debate revolves around the standardization and consistency of Vedic Mathematics. Different sources and teachers present variations of techniques and interpretations, which can lead to inconsistencies and confusion. Critics argue that without a standardized set of techniques and rules, it becomes challenging to establish a coherent and widely accepted framework for Vedic Mathematics.

Integration with Traditional Mathematics: Some mathematicians question the need for Vedic Mathematics as a separate system, advocating for a more integrated approach that combines traditional mathematics with mental calculation techniques. They suggest that mental calculation skills can be developed within the existing mathematical curriculum without the need for a separate framework like Vedic Mathematics.

It is important to engage in critical discussions and consider various viewpoints when exploring the merits and limitations of Vedic Mathematics. The debates surrounding Vedic Mathematics highlight the need for a thoughtful and balanced approach, where the strengths and weaknesses of the system are critically evaluated within the broader context of mathematical education and problem-solving.

Chapter VII
Future Prospects and Further Study

A. Emerging Research and Developments in Vedic Mathematics

While Vedic Mathematics has been a subject of study and practice for many years, there has been relatively limited recent research specifically focused on it. However, there have been some emerging developments and research interests related to Vedic Mathematics. Here are a few areas that have gained attention:

Pedagogical Approaches: Researchers are exploring the effectiveness of incorporating Vedic Mathematics techniques in educational settings. Studies are being conducted to assess the impact of Vedic Mathematics on students' mathematical abilities, problem-solving skills, and mental agility. Researchers are also examining the potential benefits of integrating Vedic Mathematics into existing mathematics curricula.

Cognitive Science Perspectives: Cognitive scientists have shown interest in studying the cognitive processes involved in Vedic Mathematics techniques. Research in this area seeks to understand the mental strategies and cognitive mechanisms employed during mental calculations using Vedic Mathematics. This research can shed light on the cognitive advantages and limitations of Vedic Mathematics techniques.

Neuroimaging Studies: There is emerging interest in investigating the neural correlates of mental calculations performed using Vedic Mathematics techniques. Neuroimaging studies, such as functional magnetic resonance imaging (fMRI), aim to identify brain regions and networks involved in mental calculations and explore potential differences between Vedic Mathematics techniques and traditional mathematical approaches.

Computational Investigations: Some researchers are exploring the computational aspects of Vedic Mathematics techniques. They aim to analyze the efficiency and computational complexity of Vedic Mathematics algorithms in comparison to traditional algorithms for specific mathematical operations. Computational studies can provide insights into the potential advantages and limitations of Vedic Mathematics in terms of computational efficiency.

It is worth noting that research in these areas is relatively limited compared to other branches of mathematics. The majority of existing research focuses on exploring the pedagogical aspects and educational implications of Vedic Mathematics. As Vedic Mathematics continues to gain attention and interest, it is possible that more research will be conducted to further investigate its theoretical foundations, applications, and potential contributions to the field of mathematics education.

B. Potential Areas of Growth and Innovation

While Vedic Mathematics has been around for centuries, there are potential areas of growth and innovation that could be explored within its framework. Here are some potential areas for growth and innovation in Vedic Mathematics:

Further Research and Validation: Conducting rigorous scientific research to validate and explore the effectiveness of Vedic Mathematics techniques in various domains, such as education, mental cognition, and problem-solving. This would involve conducting controlled studies, collecting empirical data, and comparing the outcomes with traditional mathematical approaches.

Development of Comprehensive Curriculum: Designing a comprehensive curriculum that integrates Vedic Mathematics techniques into existing mathematics education. This curriculum could be tailored for different age groups and educational levels, providing a structured and systematic approach to learning and applying Vedic Mathematics principles.

Digital Tools and Mobile Applications: Developing digital tools and mobile applications that leverage Vedic Mathematics techniques to aid in mental calculations, problem-solving, and mathematical exploration. These tools could provide interactive tutorials, practice exercises, and real-time feedback to enhance learning and application of Vedic Mathematics principles.

Online Learning Platforms: Establishing online platforms dedicated to Vedic Mathematics education, offering courses, resources, and interactive learning materials. These platforms could provide accessible and scalable avenues for individuals to learn Vedic Mathematics techniques at their own pace and convenience.

Integration with Artificial Intelligence: Exploring the integration of Vedic Mathematics principles with artificial intelligence (AI) algorithms to develop intelligent computational systems. This could involve leveraging the pattern recognition capabilities of Vedic Mathematics to enhance AI algorithms' efficiency and accuracy in mathematical calculations and problem-solving tasks.

Application in Data Science and Analytics: Exploring the potential applications of Vedic Mathematics techniques in data science and analytics. The pattern recognition and mental calculation abilities of Vedic Mathematics could be beneficial in analyzing large datasets, identifying trends, and making data-driven decisions.

Customization for Specific Domains: Adapting Vedic Mathematics techniques to specific domains or disciplines, such as engineering, finance, or cryptography. Customizing Vedic Mathematics principles to address the unique computational requirements and problem-solving challenges in these fields could provide specialized tools and approaches.

It is important to note that exploring these areas would require a balance between preserving the authenticity of Vedic Mathematics principles and embracing modern pedagogical and technological advancements. Collaborations between mathematicians, educators, researchers, and technology experts could drive innovation and open new avenues for growth within the field of Vedic Mathematics.

Chapter VIII
Conclusion

A. Summary of Key Takeaways From the Book

Key Takeaways from the book on Vedic Mathematics:

Origins and Philosophy: Vedic Mathematics is inspired by ancient Indian mathematical principles found in Vedic literature. It emphasizes mental calculation, pattern recognition, and simplification techniques.

Sutras: The 16 Sutras are fundamental principles of Vedic Mathematics that provide efficient techniques for various mathematical operations, such as addition, subtraction, multiplication, and division.

Mental Calculation: Vedic Mathematics techniques enable fast and accurate mental calculations, making it useful in everyday life, competitive exams, and certain professional fields.

Problem-Solving Skills: Vedic Mathematics promotes pattern recognition, logical reasoning, and innovative problem-solving approaches. It encourages thinking outside the box and finding creative solutions to mathematical problems.

Pedagogical Applications: Vedic Mathematics can be integrated into educational settings to enhance students' mathematical abilities, mental agility, and problem-solving skills. It offers an alternative approach to traditional mathematics education.

Practical Applications: Vedic Mathematics finds applications in various real-life scenarios, including finance, business, data analysis, architecture, and cryptography. It can aid in quick calculations, estimation, and decision-making processes.

Limitations and Criticisms: Vedic Mathematics is not a comprehensive mathematical system and does not encompass all branches of mathematics. Its historical authenticity, scope, and standardization have been subject to debates and varying viewpoints.

Remember that this summary provides a brief overview of the key takeaways, and the complete book would contain more detailed explanations, examples, and applications of Vedic Mathematics.

B. Encouraging Readers to Delve Deeper Into Vedic Mathematics

Delving deeper into Vedic Mathematics can be a rewarding journey that expands your mathematical skills and problem-solving abilities. Here are a few reasons to encourage further exploration:

Enhance Mental Calculation Skills: Vedic Mathematics offers techniques that can significantly improve your mental calculation speed and accuracy. By diving deeper into Vedic Mathematics, you can master advanced techniques and broaden your repertoire of mental calculation strategies.

Discover Hidden Patterns and Shortcuts: Vedic Mathematics is built on the principle of pattern recognition. By exploring the subject in more depth, you'll uncover intricate patterns, connections, and shortcuts that can simplify complex calculations and problem-solving.

Develop a Problem-Solving Mindset: Vedic Mathematics encourages a holistic and innovative approach to problem-solving. By delving deeper, you'll deepen your understanding of the underlying principles and develop a problem-solving mindset that extends beyond calculations, benefiting you in various mathematical and real-world scenarios.

Explore Connections with Other Mathematical Fields: While Vedic Mathematics primarily focuses on arithmetic and algebraic techniques, further exploration can reveal connections with other branches of mathematics. You may discover how Vedic principles can be applied in geometry, number theory, calculus, and even advanced mathematical concepts.

Apply Vedic Mathematics in Practical Scenarios: By delving deeper into Vedic Mathematics, you'll gain a more nuanced understanding of its applications in practical scenarios. This knowledge can be applied to various real-life situations, such as finance, business, architecture, and cryptography, enabling you to make quick and accurate calculations.

Engage in Academic Discussions and Research: Vedic Mathematics remains a subject of ongoing academic interest and research. By delving deeper, you can contribute to the scholarly discussions, explore research opportunities, and even challenge existing theories and interpretations. Engaging with academic communities can foster intellectual growth and broaden your perspective.

Remember, delving deeper into any mathematical discipline requires dedication, patience, and a thirst for knowledge. Utilize books, online resources, research papers, and engage with experts in the field to deepen your understanding of Vedic Mathematics. Embrace the challenges and enjoy the journey as you unlock the beauty and power of Vedic Mathematics.

C. Final Thoughts on the Enduring Relevance of this Ancient Mathematical System

Vedic Mathematics, despite its ancient origins, continues to hold enduring relevance in the modern world. Its principles and techniques offer valuable insights and approaches to mathematical calculations, problem-solving, and mental agility. Here are some final thoughts on the enduring relevance of this ancient mathematical system:

Mental Calculation Skills: In an age where calculators and digital devices are readily available, the ability to perform quick and accurate mental calculations remains invaluable. Vedic Mathematics equips individuals with mental calculation techniques that enhance their computational skills and foster mental agility.

Efficiency and Speed: The emphasis on efficiency and speed in Vedic Mathematics makes it applicable in various domains, including competitive exams, business, finance, and day-to-day calculations. The techniques enable individuals to save time and make rapid calculations, enhancing productivity and decision-making.

Problem-Solving Strategies: Vedic Mathematics promotes innovative problem-solving strategies, encouraging individuals to think beyond conventional approaches. The focus on pattern recognition, logical reasoning, and simplification techniques fosters a creative mindset that can be applied not only to mathematics but also to other fields requiring analytical thinking.

Educational Benefits: Incorporating Vedic Mathematics into education can yield benefits for students. It provides alternative methods for understanding mathematical concepts, engages students through interactive and mental calculation activities, and enhances their overall mathematical abilities and confidence.

Cultural Heritage: Vedic Mathematics represents a part of ancient Indian cultural heritage. Exploring and preserving this knowledge helps to appreciate the rich mathematical traditions of the past and fosters a deeper understanding of different mathematical systems worldwide.

Bridge between Ancient Wisdom and Modern Mathematics: Studying Vedic Mathematics offers an opportunity to bridge the gap between ancient wisdom and modern mathematics. It encourages individuals to delve into historical mathematical texts, explore connections with contemporary mathematical concepts, and appreciate the diverse approaches to mathematical thinking.

While acknowledging the enduring relevance of Vedic Mathematics, it is important to recognize its limitations and maintain a balanced perspective. Vedic Mathematics should be seen as a complementary tool rather than a replacement for traditional mathematics. Integrating its principles with a comprehensive understanding of mathematical concepts can enhance one's mathematical abilities and problem-solving skills.

Overall, the enduring relevance of Vedic Mathematics lies in its ability to provide alternative perspectives, mental calculation techniques, and problem-solving strategies that continue to benefit individuals in various aspects of life, education, and professional fields.

www.ingramcontent.com/pod-product-compliance
Lightning Source LLC
Chambersburg PA
CBHW022346290526
45786CB00014B/2513